Abstract

The languages and logical formalisms developed by information scientists and logicians concentrate on the theory of languages and logical theorem proving. These languages, when used by domain experts to represent their domain of discourse, most often have issues related to the level of expressiveness of the languages and need specific extensions. In this paper we analyze the levels of logical formalisms and expressivity requirements for the development of ontologies for manufacturing products. We first discuss why the representation of a product model that needs to be shared across globally networked enterprises is inherently complex and prone to inconsistencies. We then explore how these issues can be overcome through a structured knowledge representation model. We report our evaluation of OWL-DL (Ontology Web Language-Description Logic) in terms of expressivity and of the use of SWRL (Semantic Web Rule Language) for representing domain-specific rules. We present a case study of product assembly to document this evaluation and further show how the OWL-DL reasoner together with the rule engine can enable reasoning of the product ontology. We discuss how the proposed product ontology can be used within a manufacturing context.

Table of Contents

1. Introduction 1
2. DL expressivity in product modeling 3
   2.1 DL for information representation 3
   2.2 DL for inference mechanisms 3
   2.3 Domain-specific rules 4
3. Languages and tools 5
   3.1 Modeling Languages 5
   3.2 Modeling Tools 6
4. Description of the information model 7
   4.1 Core Product Model 7
   4.2 Open Assembly Model 7
5. Description of the model inference mechanisms 11
   5.1 Description logic 11
   5.2 Rule-Based Inference 13
6. Product Model Example 16
   6.1 Example of DL reasoning 17
   6.2 Example of rule-based reasoning 18
   6.3 Example of combining DL and rule-based reasonings 19
7. Usage scenario of the product model in manufacturing 20
8. Conclusions 23
9. Disclaimer 24
10. References 25

NISTIR 7481

An Evaluation of Description Logic for the Development of Product Models

Xenia Fiorentini
Sudarsan Rachuri
Mahesh Mani
Steven J. Fenves
Ram D. Sriram
Manufacturing Systems Integration Division
Manufacturing Engineering Laboratory

APRIL 2008

U.S. Department of Commerce
Carlos M. Gutierrez, Secretary

National Institute of Standards and Technology
James M. Turner, Director

List of Figures

Figure 1: Connection between the OWL ontology and SWRL rules 6
Figure 2: Core Product Model 9
Figure 3: Open Assembly Model 10
Figure 4: Example of a case where rules are needed 13
Figure 5: Example of a property rule 14
Figure 6: Example of an association rule 14
Figure 7: Example of a partOf rule 15
Figure 8: Example of an acyclic rule 15
Figure 9: Case study: planetary gear system 16
Figure 10: Manufacturing Information flow. Adopted/modified from [21] 20
Figure 11: Prismatic part example 21
Figure 12: Sequence of drilling operations in OWL-DL 22

List of Tables

Table 1: Examples of inference mechanisms 3
Table 2: Examples of DL expressivity in product modeling 11
Table 3: Example of DL reasoning 17
Table 4: Example of rule-based reasoning 18
Table 5: Rules needed to connect Assembly with its ArtifactAssociations 19
Table 6: Example of combining DL and rule-based reasonings 19

1 Introduction

In a typical industrial scenario a number of organizational entities collaborate to accomplish various tasks by sharing resources, applications and services throughout the product lifecycle. The immediate issue is the interoperability of these resources, applications and services. Interoperability requires a common high level and interoperable model of the product across the extended and networked enterprise.

The development of such a high level interoperable model poses many challenges, for example: i) complex nature of interactions in product modeling and ii) representation of the information content and abstraction principles used.

The information model for representing mechanical assemblies (products) is inherently complex owing to:

- The variety of information elements to be represented: function, behavior, structure, geometry and material, assembly features, tolerances and various levels of interaction of these concepts,
- The abstraction principles needed to represent the information model: generalization, grouping, classification, and aggregation [1].

The following partial list sketches some of the issues with respect to each information element:

- *Function:* one aspect of what the artifact is supposed to do. The artifact satisfies the engineering requirements largely through its function [2].
- *Behavior:* information supporting the simulation of the product under some given conditions. This simulation could be, for example, kinematics, dynamics and control systems.
- *Structure:* the individual parts that constitute the assembly, the hierarchy of the composition tree (parts-subassemblies-assembly) and the associated Bill of Materials (BOM).
- *Geometry and material:* a generic shape, chosen by the designer at early stages of the lifecycle and a particular geometry captured in one or more CAD (Computer-Aided Design) models.
- *Features:* a portion of the artifact's form that has some specific function assigned to it. An artifact may have design features, analysis features, manufacturing features, etc., as determined by their respective functions [2].
- *Tolerances:* tolerance design is the process of deriving a description of geometric tolerance specifications for a product from a given set of desired properties of the product. Tolerancing includes both tolerance analysis and tolerance synthesis [2].

The second source of complexity is due to the abstraction principles needed to represent the information on products. The model may incorporate the following mechanisms:

- *Generalization versus specialization:* relationships built through intentional properties, e.g., "gear shifts are mechanical assemblies." This abstraction principle involves a hierarchical mechanism where concepts are categorized through the general knowledge of the problem.
- *Grouping versus individualization:* relationships built through extensional properties, e.g., "manual gear shifts are gear shifts." In this case concepts are categorized through the specific knowledge of the represented domain and the group can even be not homogeneous meaning that the group can contain disparate things.
- *Classification versus instantiation:* relationships between a real object (individual) and the concept it belongs to, e.g., "my gear shift is a gear shift." In modeling, particular attention needs to be paid to the establishment of the boundary between concepts and real objects.
- *Aggregation versus decomposition:* part-of relationships between an element and its constituents, e.g., "gear shifts are part-of cars" and "my gear shift is part-of my car."

Other types of mechanisms may be required, for example, a given chemical compound "consists of" many other chemicals. For more details regarding general part-whole formalisms readers can refer to the General Extensional Mereology [3].

This paper outlines a method for evaluating the appropriate level of expressiveness to capture both the information content and the abstraction principles discussed above, with the aim of developing a consistent formal model for product assemblies. We use the terms expressiveness to mean both language expressiveness and processable expressiveness [4]. The language expressiveness is related to the language symbols, rules, conventions and vocabulary while the processable expressiveness is related to the computability. For more detailed discussion on the issues related to the computational complexity, please refer to [5] [6].

The procedure one has to follow to represent product models can be summarized as follows:
1. Select Description Logic (DL).
2. Select a language that well support DL.
3. Evaluate extensions (to incorporate domain specific rules) to the language and pick the appropriate rule language.
4. Build the model and evaluate inference mechanisms.

We believe that this study may aid in understanding the considerations involved in choosing appropriate logical frameworks for product ontologies.

This paper is structured as follows. We first describe how the expressivity of DL can help in developing a consistent formal model for product assemblies. We then find the language that captures most of the expressivity of DL. We use that language to build a product model and to show how the DL expressivity is used in the product representation. For practical purposes, we finally illustrate how the resulting product model can be used within a design context and a manufacturing context.

2 DL expressivity in product modeling

Description Logic is a family of knowledge representation languages used to represent the knowledge of a domain in a structured fashion. The domain is modeled by means of concepts and roles, which denote, respectively, classes of objects and relationships between objects. The concepts and roles, together with knowledge specification mechanisms, form the knowledge base. Automatic reasoning procedures can be performed on the knowledge base.

DL is decidable, that is, there exists an automatic reasoning procedure such that, for every knowledge specification mechanism in the logic, the reasoning procedure is capable of deciding whether the mechanism is valid or not [7].

We have to choose the level of expressiveness needed to represent the product information content and to include the abstraction principles needed to represent it. Expressiveness should enable explicit information representation (product model) and support inference mechanisms, i.e., mechanisms to find implicit consequences based on the explicit information.

2.1 DL for information representation

The DL formalism allows us to create a concept level hierarchy of the knowledge using is-a relationships (e.g., car is-a vehicle), to express complex roles (properties) between concepts (e.g., cars have exactly four wheels while bicycles have exactly two wheels) and to declare the membership of an individual in a concept (e.g., myCar belongs to the concept of cars).

2.2 DL for inference mechanisms

For inference mechanisms, consider the examples in Table 1. In the second column we use the concepts of *Vehicle*, *Car*, *Bicycle* and *Wheel* to create our knowledge base and to query it. In the third column we present the DL mechanisms that allow for answers to those queries. In the fourth column we show answers to those queries.

Table 1: Examples of inference mechanisms

	Question	DL mechanisms	Answer
1	We subsume the concept of *Car*[1] in the concept of *Bicycles*[2]. Is it logically correct?	The "consistency checking" mechanism finds whether a concept admits at least one individual.	No, the model is inconsistent. There cannot be an individual that has four wheels and is a bicycle at the same time.
2	We introduce the concept of *ElectricCar*. What is its position in the hierarchy?	The "subsumption" mechanism finds implicit sub-concept relationships.	In the concept's hierarchy, the *ElectricCar* concept is a sub-concept of *Car*.

3	We declare myCar as an individual of the concept of *Vehicle* with four wheels. Is it a *Car* or a *Bicycle*?	The "realization reasoning" mechanisms finds the most specific concept for each individual.	myCar has four wheels, so it is an instance of the concept of *Car*.
4	Which cars have the same kind of wheels?	The "retrieval" mechanism finds the individuals that are instances of a given concept or intersection of concepts.	The set of different instances of *Car* that have matched wheels.
5	We declare a wheel part-of a car but the engine powering that wheel is part-of another car. Is it logically correct?	DL can not help here since the roles do not pertain to concepts but particular individuals.	The individual of the *Wheel* concept is connected to the wrong individual of *Car*. The part-of property between the concepts of *Wheel* and *Car* is still correct.

[1] *Vehicle* with four wheels
[2] *Vehicle* with two wheels

The DL formalism consists of four reasoning mechanisms [7]: consistency checking, subsumption, realization, and retrieval. Each of them provides the answer for one of the first four questions. The fifth question represents a different situation, it falls outside of DL. To answer this question we need to represent appropriate role paths between instances and not between classes as in the case of the first four questions. In other words it is the role path, going from the instance of *Wheel* to the instance of *Car* passing through the instance of *Engine*, which has to be constrained. To answer the fifth question, we have to introduce in the representation new elements: domain-specific rules.

2.3 Domain-specific rules

Domain-specific rules are defined to add specific constraints in the knowledge base. These rules are in the form of implications between an antecedent (body) and a consequent (head): whenever the conditions specified in the antecedent hold, then the conditions specified in the consequent must also hold. These rules not only allow the declaration of the membership of an individual to a concept, but also the declaration of properties between individuals. In the fifth example given in Table 1, a rule can state that if a wheel is powered by an engine and that engine is part-of a car, then the wheel has to be part-of the same car.

In order to represent knowledge in the assembly domain, i.e., to answer all five questions in Table 1, we need to combine both DL expressivity and domain-specific rules.

3 Languages and tools

Our next goal is to find modeling languages and tools able to implement both the DL expressivity and the domain-specific rules.

3.1 Modeling Languages

The most common languages used for product modeling are:

- Unified Modeling Language (UML) [8]
- Entity-Relationship diagrams (ERD) [9]
- EXPRESS [10]
- Ontology Web Language (OWL-DL, version 1.0) [11].

In UML, the modeling elements are substantially aligned with the needs of object-oriented programming so that the correspondence with DL expressivity is low [12]. On the other hand, the expressivity of UML is embedded in its meta-modeling architecture called Meta-Object Facility (MOF). This architecture is organized in four layers, from M3 to M0, where each layer provides precise constructs and rules for creating models in the successive layers [13].

ERD was developed for the organization of information within databases, therefore the correspondence with DL expressivity is even lower than for UML [9].

In EXPRESS the correspondence with DL is not high but the expressive power is enhanced with algorithms not captured in the DL expressivity. These algorithms define the entities' behavior using functions, procedures, and rules.

Among the listed languages, OWL is the most appropriate to implement the DL constructs needed for product modeling. Each DL sublanguage is named with a combination of letters (acronyms), e.g., $\mathcal{ALC}$, $\mathcal{SHOIN}$, and $\mathcal{SHIQ}$,: each letter associates to the sublanguage its expressivity. OWL-DL is classified as $\mathcal{SHOIN}^{(D)}$. Although decidable, OWL-DL could become intractable, especially when dealing with large ontologies: for this reason, part of our effort is still focusing on its computational complexity. Its meta-modeling architecture is flat (i.e., not organized in layers) but the expressiveness is contained in the language elements themselves. These language elements have been designed with the aim of using DL for the semantic web to enable interoperability between systems through semantic data representation.

The use of the XML (Extensible Markup Language) syntax within OWL facilitates the exchange of models between agents, while the OWL features give the model the expressive power needed for ontological representation. The word "ontology" in this paper is meant as a collection of concepts on which a set of axioms is specified for performing logical inference.

3.2 Modeling Tools

We decided to test the OWL expressiveness for product modeling by building a product ontology (see Section 4) and by performing inference mechanisms on it (see Section 5). We developed the ontology in OWL-DL version 1.0, using Protégé-OWL 3.3 [14] to edit it. We excluded OWL Lite and OWL Full from consideration because of their low formal complexity and the hard computational problems, respectively. The Protégé ontology editor supports $\mathcal{SHOIN}^{(D)}$. OWL-DL provides the expressiveness of $\mathcal{SHOIN}^{(D)}$, and OWL 1.1 is based on $\mathcal{SROIQ}^{(D)}$.

For DL inference, we used the reasoning engine RACERPro [15]. We chose this reasoning engine because it is easily accessible through the OWL Plug-in in Protégé. For rule-based inference, we used the Semantic Web Rule Language (SWRL) to write the rules [16]. Since the combination SWRL and OWL-DL is undecidable, we selected only the DL-safe portion of SWRL. The rules are edited directly in Protégé-OWL through the SWRLTab, an extension to the editor, and then executed by Jess, a rule engine for the Java platform that supports rule-based programming. We used the Jess Bridge in order to:

- merge SWRL rules and relevant OWL data
- input them to the Jess engine, and
- return the new inferred information to the ontology.

Figure 1 depicts how the OWL data and the SWRL rules are connected. The bold arrows indicate the flow of information (initial data, rules and final data) while the dashed arrows indicate how the Jess Bridge enables that flow.

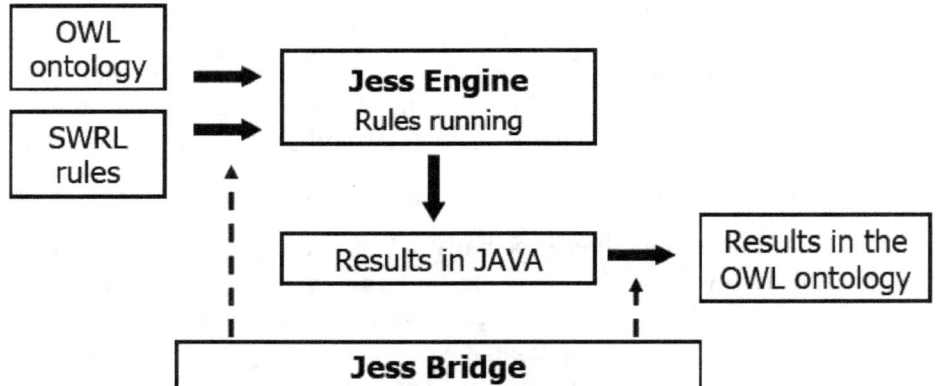

Figure 1: Connection between the OWL ontology and SWRL rules

4 Description of the information model

Core Product Model (CPM) [2] was intended to form a base for representing a product model that could respond to the demands of the next generation CAD systems besides providing improved interoperability among future software. Based on the Core Product Model and Open Assembly Model (OAM), [17] presents two ontological models. These two models were developed at the National Institute of Standards and Technology (NIST) as part of the ongoing work related to product representation for lifecycle management [2]. A brief description of these two models is given below.

4.1 Core Product Model

The model is composed of two ontologies: which are OWL versions of CPM and OAM. The concepts (classes in OWL) in CPM are grouped into four categories (see Figure 2):

- Classes that provide supporting information for the objects (abstract classes): *CoreProductModel*, *CommonCoreObject*, *CommonCoreRelationship*, *CoreProperty* and *CoreEntity*.
- Physical or conceptual objects classes: *Artifact*, *Feature*, *Port*, *Specification*, *Requirement*, *Function*, *Flow*, *Behavior*, *Form*, *Geometry* and *Material*.
- Classes that describe associations (relationships) among the objects: *Constraint*, *EntityAssociation*, *Usage* and *Trace*.
- Classes that are commonly used by other classes (utility classes): *Information*, *ProcessInformation* and *Rationale*.

The hierarchy of classes begins from *CommonCoreEntity*. This class represents real objects and relationships or associations between them. The two subclasses of *CommonCoreEntity* are *CommonCoreObject* and *CommonCoreRelationship*. *CommonCoreObject* is the parent class for all the object classes. *CommonCoreRelationship* and its specializations, the *EntityAssociation*, *Constraint*, *Usage* and *Trace* relationships, can be applied to individuals of classes derived from this class. *CommonCoreRelationship* is the base class from which all association classes are specialized. It also serves as an association to the *CommonCoreObject* class. *CoreEntity* is an abstract class from which the classes *Artifact* and *Feature* are specialized. *EntityAssociation* relationships may be applied to entities in this class. *CoreProperty* is an abstract class from which the classes *Function*, *Flow*, *Form*, and *Material* are specialized. *Constraint* relationships may be applied to individuals of this class.
For further details, please, refer to [2].

4.2 Open Assembly Model

OAM incorporates information about assembly relationships and component composition; the representation of the latter is by the class *ArtifactAssociation*, which represents the assembly relationship that generally involves two or more artifacts. *ArtifactAssociation* is specialized into the following classes: *PositionOrientation*,

Relative-Motion and *Connection*. *ArtifactAssociation* is directly connected to *Assembly* to allow the possibility to check the assembly relationship involved in the *Assembly* through the property *ArtifactAssociation2Assembly* (see Figure 3).

An assembly is a composition of its subassemblies and parts. The *Assembly* and *Part* classes are sub-classes of the CPM *Artifact* class. A *Part* is the lowest level component. Each assembly component (whether a sub-assembly or part) is made up of one or more features, represented in the model by *OAMFeature*, a subclass of the CPM *Feature* class. *OAMFeature* has tolerance information, represented by the class *Tolerance*.

The class *AssemblyFeatureAssociation* (AFA) represents the association between mating assembly features through which relevant artifacts are associated. The class *ArtifactAssociation* is the aggregation of *AssemblyFeatureAssociation*. The class *AssemblyFeatureAssociationRepresentation* (AFAR) represents the assembly relationship between two or more assembly features. This class is an aggregation of *ParametricAssemblyConstraints*, *KinematicPair*, and/or *KinematicPath* between assembly features. *KinematicPair* defines the kinematic constraints between two adjacent artifacts (links) at a joint. *KinematicPath* provides the description of the kinematic motion. For further details, please, refer to [2].

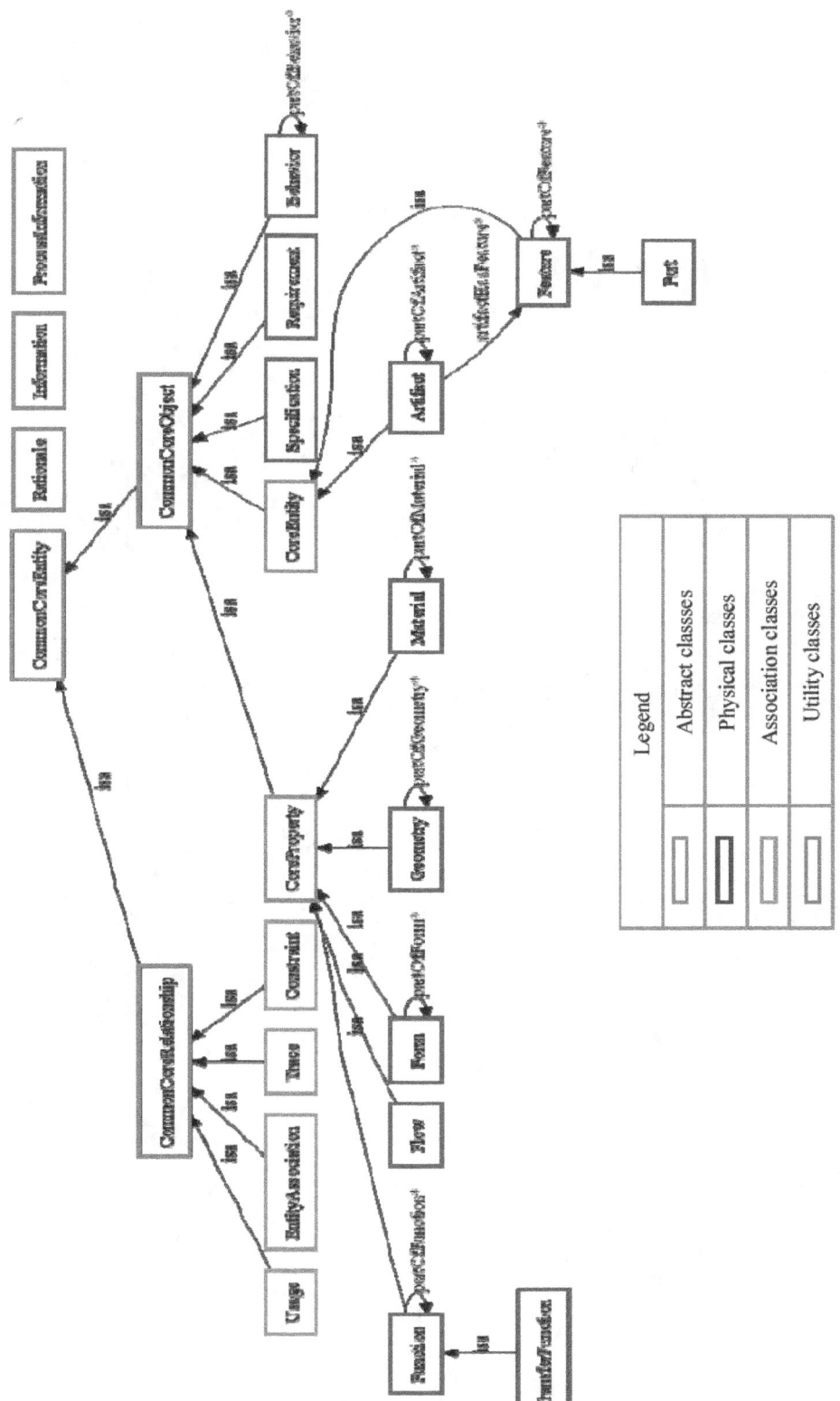

Figure 2: Core Product Model

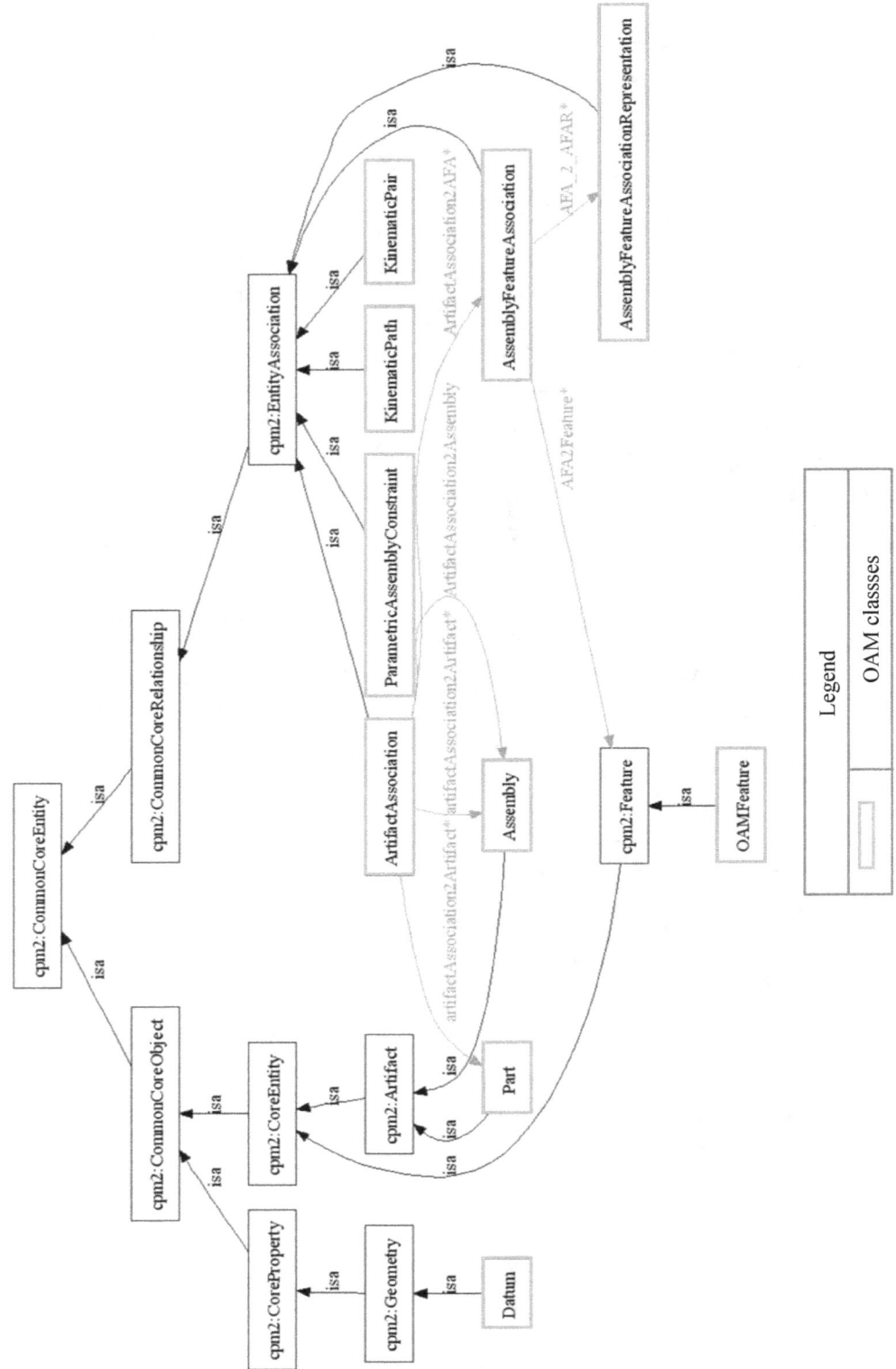

Figure 3: Open Assembly Model

5 Description of the model inference mechanisms

The main benefit of using an ontology for product modeling is the possibility of performing inference on the declared classes and individuals. In this section we will discuss the two different inference mechanisms used in our product ontology: inference based on description logic and inference based on domain specific rules

5.1 Description logic

In Table 2 we give some examples on how the DL expressivity included in OWL is used within the axioms defined in the model. In the first column of the table, the DL expressivity of OWL-DL, i.e., $\mathcal{SHOIN}^{(D)}$, is divided according to the DL notation [7].

The third column in Table 2 indicates the expressivity associated with each letter of the DL notation.

Table 2: Examples of DL expressivity in product modeling

Notation	No	Expressivity	Description	Examples of Axioms using the Expressivity
$\mathcal{AL}$	1	Universal concept	The concept that contains all the individuals.	The concept of *Thing*: included in every OWL ontology.
	2	Bottom concept	The concept without any individual.	The concept of *Nothing*: included in every OWL ontology.
	3	Atomic concept	A concept name.	The concept of *Assembly*.
	4	Atomic negation	The negation of an atomic concept.	The concept of *Part* consists of those individuals that are not *Assemblies*.
	5	Value restriction	All the individuals that are in the relationship with the described concept belong to a specified concept.	A *Feature* can be connected through the property Feature2ParametricAssemblyConstraint only to the concept *ParametricAssemblyConstraint*.
	6	Intersection of concepts	The set of individuals belonging to both the concepts.	An *OAMFeature* is the intersection between the concept of *Feature* and the concept of the individuals connected at least with one *AFA* through the property feature2AFA. An *OAMFeature* can be automatically recognized by giving the definition of that concept.
$\mathcal{S}$	7	Transitive properties	For all individuals a, b, and c, if a is related to b and b is related to c, then a is related to c.	The property artifactHasPart, used to connect an *Assembly* with all its components, is transitive. Since for transitive properties it is impossible to specify cardinalities, the model includes also the property artifactHasPart_direct to connect the *Assembly* to its direct subassemblies or *Parts*.

Notation	No	Expressivity	Description	Examples of Axioms using the Expressivity
$\mathcal{H}$	8	Role hierarchy	If P1 is a subproperty of P2, then the property extension of P1 (a set of pairs) should be a subset of the property extension of P2 (also a set of pairs).	The property artifactHasPart_direct is a subproperty of the property artifactHasPart. When the direct property holds, the indirect one holds as well.
$\mathcal{O}$	9	Enumerated classes	The concept is made of exactly the enumerated individuals.	Two *CommonCoreObject* can be linked through the *CommonCoreRelationships* "AlternativeOf", "IsSameAs", "VersionOf", "IsBasedOn", "DerivedFrom": these are the enumerated individuals of the range class of the link.
$\mathcal{I}$	10	Inverse properties	For all individuals a and b, iff a is related to b, then b is related to a through the inverse property.	The property partOf is the inverse of artifactHasPart. When an *Assembly* is connected to its component through artifactHasPart, the component will be connected to the *Assembly* through partOf.
$\mathcal{N}$	11	Cardinality restrictions	The concept is constrained to have a number of values of a particular property.	An *ArtifactAssociation* has to link at least 2 *Artifacts*. An inconsistency will be identified if not.
$\mathcal{F}$	12	Functional properties	The individuals of certain concepts have unique property fillers for a given property.	A *KinematicPair* can be referred only to one *AFAR*.
$\mathcal{E}$	13	Full existential quantification	The set of all individuals in the domain which has at least one specified R-successor.	A *DatumFeature* has to have some connections with *AssemblyFeatureAssociation*. If it doesn't the reasoner will recognize an inconsistency.
$\mathcal{U}$	14	Concept union (disjunction)	The set of the individuals belonging at least to one of the disjointed concepts.	The property that connects *ArtifactAssociation* to the assembled components has as range the concept union of *Assembly* and *Part*.
$(\mathcal{D})$	15	Datatype properties	Property for which the value is a data literal, such as a string or a number.	*CommonCoreEntities* have names: the property links *CommonCoreEntity* to a string.

All OWL constructs ($\mathcal{SHOIN}^{(\mathcal{D})}$) are used in OAM but only the inference mechanisms of $\mathcal{SHIN}^{(\mathcal{D})}$ are performed since the reasoning capability of RacerPro in the presence of enumerated classes ($\mathcal{O}$) is incomplete [18]. In our ongoing research we are evaluating the Pellet reasoner ($\mathcal{SHOIQ}^{(\mathcal{D})}$) because it is more compatible with OWL 1.1 ($\mathcal{SROIQ}^{(\mathcal{D})}$), the newest version of OWL [19].

Following the axioms, some examples are given in Table 2. With this set of axioms the reasoner is able to:

- Query and search the model.
- Check its consistency.
- Perform inference on the classes' hierarchy.
- Perform inference on the membership of the individuals to the classes.

In OWL-DL 1.0 a property is declared in terms of its domain, range and characteristics such as transitivity or reflexivity. In cases where it is required to impose specific conditions or restrictions we need to specify them using some rules. For example, if class *Car* and class *Person* are connected through the property hasOwner and class *Person* and class *Garage* are connected through the property isRenter and class *Car* and class *Garage* are connected through the property isParked (see Figure 4) then to infer that a particular person's car is parked in the garage the person rents we need a rule to specify this explicitly.

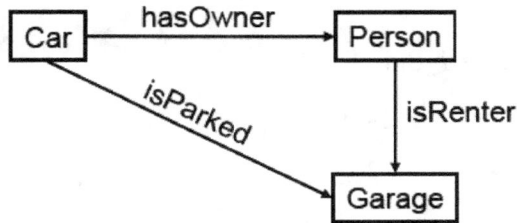

Figure 4: Example of a case where rules are needed

In OWL 1.1 the above rule can also be achieved through property chains but in our opinion not in all cases the rules can be replaced by property chains.

5.2 Rule-Based Inference

We use SWRL [16] rules in order to:

- associate individuals to new classes: we use this capability to associate an individual to a class creating inconsistencies in the ontology
- create properties between individuals.

We classify these rules into four groups:
- property rules
- association rules
- partOf rules, and
- acyclic rules.

In the diagrams representing the rules (Figures from 5 to 8), we use rectangles to identify classes and ovals to identify individuals.

Property rules create new properties between individuals once some other properties are declared. The property rules incorporate the meaning into the ontology. For example, the Jess engine associates the *ArtifactAssociation* directly to the *Assembly* once the structure of the *Assembly* and the *ArtifactAssociation* between its subassembly are declared. Figure 5 shows how the rule connects the master Assembly 1 to the ArtifactAssociations existing between its subcomponents Assembly 2, Part 1, and Assembly 3. The legend in Figure 5 is shared also by Figure 6,7 and 8.

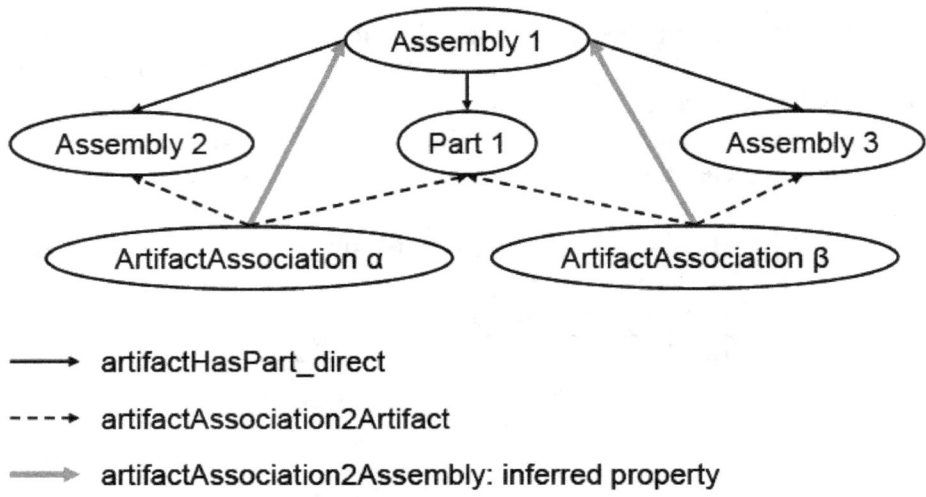

Figure 5: Example of a property rule

Association rules represent the binary relationships between association classes and object classes (see Figure 6). A minimum cardinality 2 is applied in the OWL model, and then a SWRL rule specifies that if two different individuals of the association class are connected to the same individuals of the object class, then these two association individuals are the same (sameAs). In this way a unique *ArtifactAssociation* can be connected to the same individuals of *Artifact*.

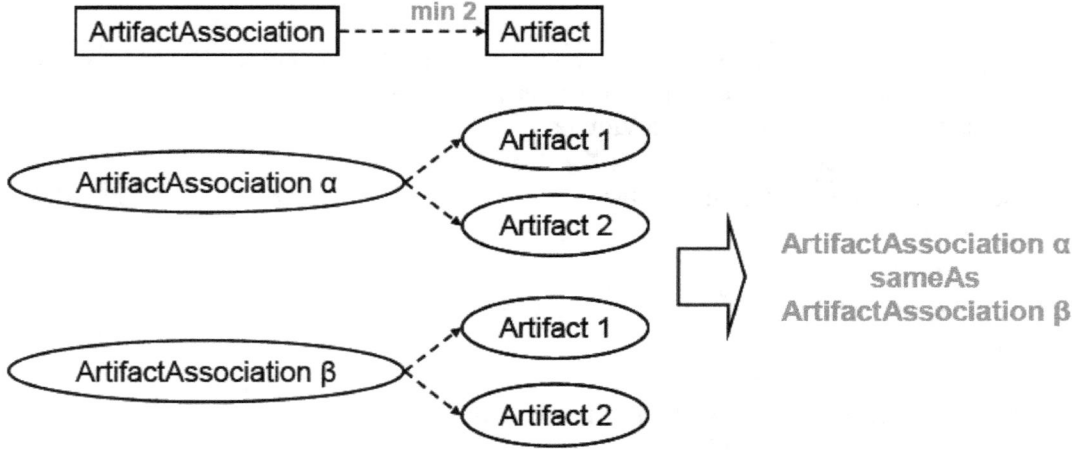

Figure 6: Example of an association rule

PartOf rules create the right structure of assembly, i.e., enable the assemblies to distinguish between the direct and indirect part-of properties. After executing the partOf rules, the indirect property links an assembly with all its parts (example in Figure 7).

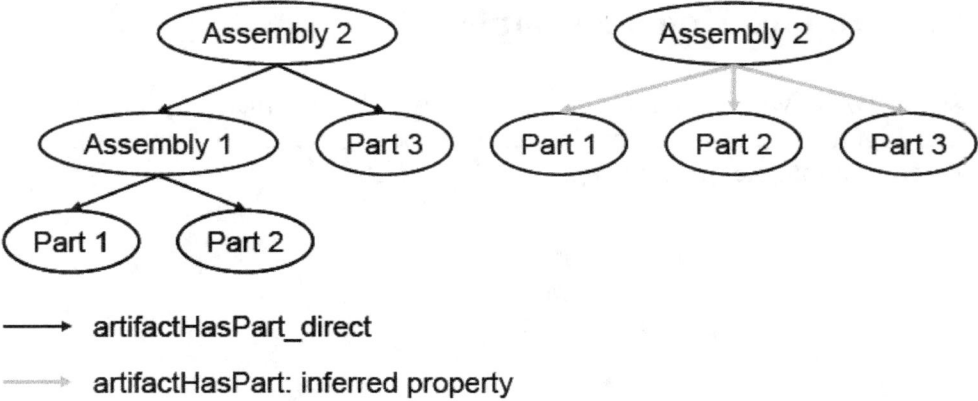

→ artifactHasPart_direct

⇢ artifactHasPart: inferred property

Figure 7: Example of a partOf rule

Acyclic rules instantiate classes of the kind *not-allowed*, to identify the individuals that, although declared, are included in a part-of cycle. Since no inference mechanism can delete wrong information from the ontology, we insert the wrong information in the not-allowed classes through the acyclic rules. Since the not-allowed classes are declared disjoint from the original ones, the reasoner will detect an inconsistency.

Take the example in Figure 8: the assembly 2 is composed by itself (assembly 2 is composed by assembly 1 that is in turn composed by assembly 2).

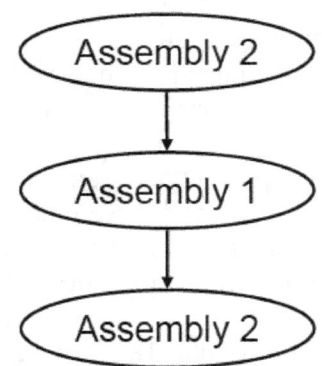

Figure 8: Example of an acyclic rule

In this example, both Assembly 2 and 1 are individuals of the same class, so no axioms can be applied to the relationships between them. For this reason, we create the *NotAllowedAssembly* class and instantiate it through the acyclic rules. Since the classes *NotAllowedAssembly* and *Assembly* are disjoint, the reasoner detects an inconsistency.

We give examples of the reasoning mechanisms employed in the next section, where a case study is presented for the exploration of the potentialities of the ontology assembly representation.

6 Product Model Example

To test the OAM with reasoning capabilities, we chose a planetary gear system as an example. Figure 9 and the summary presented below are taken from [20]. For a more detailed description, please refer to [2].

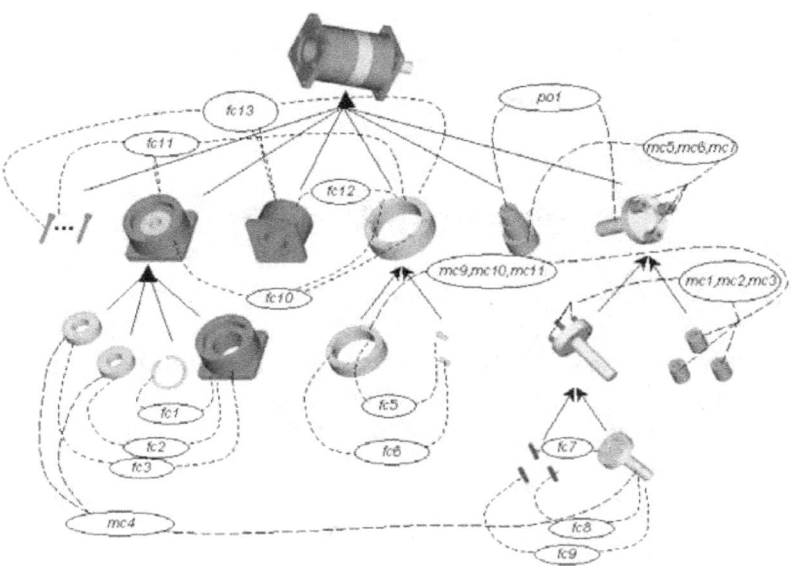

Figure 9: Case study: planetary gear system

The planetary gear system is composed of two parts and three sub-assemblies. The parts include the input-housing and the sungear. The three subassemblies include: (1) the output end assembly comprising two bearings, a washer, and the output housing; (2) the ring gear assembly comprising a ring gear and two ring-gear pins; and (3) the planet gear holder assembly comprising three planet gears and a planet carrier assembly, which further decomposes into the output shaft and three planet-gear pins. In total there are 30 different parts. The connections and pairs between different artifacts are of different types: fixed connection (fc), movable connection (mc) or position orientation (po).

To represent the use case we declare in total 187 individuals and 277 properties between the individuals. These individuals comprise not only the *Artifacts* but also their *Features*, their *Geometries*, their *Tolerances* and their connections through the association classes. Out of the 187 individuals, 70 are declared to belong to the class *Thing*, parent of all the classes in the ontology. The reasoner, using the classes and properties axioms, classifies these 70 individuals into their proper classes.

The inference mechanisms concern not only the individuals in the ontology but also the properties between the individuals. The editor Protégé-OWL automatically defines the inverse and the parent properties. Since all the properties in the model have their inverse, the editor defines 277 inverse properties, one for each declared direct property. Moreover, the editor defines all the properties parents of the asserted properties.

After performing the DL reasoner, the rule-based inference found additional 170 properties that are added to the ontology.

In the following sections, we provide three examples of the inference mechanisms we used: the first is based on description logic, the second is based on domain-specific rules while the third combines both of the previous ones.

6.1 Example of DL reasoning

Table 3 presents an example of description logic reasoning from the case study. The class *Artifact* and its subclasses are the main focus. In OAM we describe the class *Part* with a necessary and sufficient condition: *Parts* are *Artifact*s without subassemblies. In other words, in DL expressivity, the concept of *Part* is the intersection between the concept of *Artifact* and the concept of the *Thing* having cardinality 0 on the property artifactHasPart_direct ($\mathcal{AL}$ expressivity, number 6 in Table 2). This property has as domain (the class owning the property) and as range (the class of the values of the property) the class Artifact.

Moreover, we describe the class *Assembly* with a necessary condition: *Assemblies* must have at least two *Artifacts* connected through the inherited property artifactHasPart_direct. In other words, in DL expressivity, we apply a cardinality restriction ($\mathcal{N}$ expressivity, number 11 in Table 2) applied to the concept of *Assembly*.

We then define *Assembly* and *Part* as partitions of the class *Artifact*, i.e., the concept of *Artifact* is made by the union ($\mathcal{U}$ expressivity, number 14 in Table 2) of the disjoint concepts *Assembly* and *Part*. As a result, an individual of *Artifact* (Planet_Carrier_Assembly in this example) composed by other *Artifacts* (Output_Shaft, Planet_Gear_Pin_1, Planet_Gear_Pin_2, Planet_Gear_Pin_3) is inferred to be an individual of *Assembly*.

Table 3: Example of DL reasoning

AIM	Infer that an *Artifact* composed by other *Artifacts* is an *Assembly*
CLASSES	*Artifact, Assembly*
PROPERTIES	artifactHasPart_direct (Range: *Artifact*, Domain: *Artifact*)
RESTRICTION	On *Assembly:* artifactHasPart_direct min 2 (an *Artifact* is an *Assembly* only if it is related with at least 2 other *Artifacts*)
INPUT	An individual of *Artifact* (Planet_Carrier_Assembly) is composed through artifactHasPart_direct by 4 individuals of *Part* (Output_Shaft, Planet_Gear_Pin_1, Planet_Gear_Pin_2, Planet_Gear_Pin_3)
OUTPUT	Planet_Carrier_Assembly is reclassified as an individual of the class *Assembly*

6.2 Example of rule-based reasoning

Table 4 presents an example of rule-based reasoning from the case study: the structure of an Assembly is described by its parts/subassemblies and by the relationship between its components.

Table 4: Example of rule-based reasoning

AIM	Infer the relation between *Assembly* and *ArtifactAssociation*
CLASSES	*Assembly, Part, ArtifactAssociation*
PROPERTIES	artifactAssociation2Assembly (Range: *ArtifactAssociation*, Domain: *Assembly*)
RULES	If the components of an *Assembly* are linked through an *ArtifactAssociation*, then relate that *ArtifactAssociation* to the *Assembly* (see Table 5)
INPUT	An individual of *Assembly* (Output_Housing_Assembly) is composed of Bearing_1, Bearing_2, Output_Housing and Washer through artifactHasPart_direct. These individuals are connected with individuals of the class *ArtifactAssociation*
OUTPUT	Output_Housing_Assembly is linked with the corresponding individuals of ArtifactAssociation (fc_1, fc_2, fc_3, mc_4) through the ArtifactAssociation2Assembly property.

In this example, Output_Housing_Assembly is composed by Bearing_1, Bearing_2, Output_Housing and Washer. The ArtifactAssociations connect Washer with Output_Housing (fc_1), Bearing_1 with OutputHousing (fc_2), Bearing_2 with Output_Housing (fc_3) and Bearing_1 with Bearing_2 (mc_4).

The aim of the reasoning is to correctly relate the Output_Housing_Assembly to the ArtifactAssociations involved in the assembly. In this case we can not use OWL declarations since the condition for creating the new relation is dependent on the specific properties each individual possesses. For this reason we have to resort to SWRL rules.

In this example we need four different property rules (see Table 5). Each of them takes into account a different scenario:

- *Rule 1* is applied when the description of the *Assembly* is detailed (the *AssemblyAssociation* connects two or more *Parts*) and the *Assembly* has at least one subassembly that is a *Part*. The antecedent of the rule indicates that one *Part* is directly part-of the *Assembly* while the other *Part* is indirectly connected to the *Assembly*.
- *Rule 2* is applied when the description is detailed but the *ArtifactAssociation* exists between *Parts* that are not directly subassemblies of the *Assembly*. This means that the *Assembly* is composed by other subassemblies and each subassembly has a *Part* involved in the *Assembly*. In the antecedent of Rule 2 we explore the indirect property to search these *Parts* in the subassemblies.
- *Rule 3* is applied when the description is not detailed so that the *Assembly* is composed by two or more subassemblies connected together.

- *Rule 4* is similar to the third but is useful when the *Assembly* is made by a *Part* and a subassembly.

Table 5: Rules needed to connect Assembly with its ArtifactAssociations

Rule 1	Rule 2	Rule 3	Rule 4
artifactHasPart_direct(?x, ?y) Part(?y) artifactHasPart_direct(?x, ?z) Part(?z) differentFrom(?y, ?z) part2AA(?y, ?a) part2AA(?z, ?a)	artifactHasPart_direct(?x, ?y) Assembly(?y) artifactHasPart_direct(?x, ?z) Assembly(?z) differentFrom(?y, ?z) artifactHasPart(?y, ?q) Part(?q) artifactHasPart(?z, ?r) Part(?r) differentFrom(?q, ?r) part2AA(?q, ?a) part2AA(?r, ?a)	artifactHasPart_direct(?x, ?y) Assembly(?y) artifactHasPart_direct(?x, ?z) Assembly(?x) part2AA(?y, ?a) part2AA(?z, ?a)	artifactHasPart_direct(?x, ?y) Assembly(?y) artifactHasPart_direct(?x, ?z) Part(?z) part2AA(?y, ?a) part2AA(?z, ?a)
→ Assembly2ArtifactAssociation(?x, ?a)			

6.3 Example of combining DL and rule-based reasonings

Table 6 presents an example of combining both DL and rule-based reasoning. The focus is the composition hierarchy of an *Assembly*. The goal in this example is to avoid cyclic composition hierarchies, i.e., hierarchies in which an assembly contains itself. Since composition hierarchies are constituted by individuals of the same class *Assembly*, we can not use any DL axiom to impose the acyclicity constraint on the hierarchy. The use of domain-specific rules is then the only solution (see Section 2).

We create in the product ontology the class *NotAllowedAssembly*, disjoint from the class *Assembly*. *NotAllowedAssembly* will contain all the individuals of the Assembly involved in a cyclic hierarchy composition (Planetary_Gear_System_Assembly in the case of Table 6). We create a SWRL rule to automatically instantiate this class. After executing the rule, the individuals involved in the cyclic hierarchy will belong to both the classes *Assembly* and *NotAllowedAssembly*. Since these two classes are declared disjoint, the DL reasoner will detect an inconsistency.

Table 6: Example of combining DL and rule-based reasonings

AIM	Infer an inconsistency in case of a cyclic composition of an *Assembly*
CLASSES	*Assembly, NotAllowedAssembly*
PROPERTIES	artifactHasPart (Range: *Artifact*, Domain: *Artifact*)
RULES	If an *Assembly* is composed of itself, then the *Assembly* will belong to the class *NotAllowedAssembly*
RESTRICTION	*Assembly* and *NotAllowedAssembly* are disjoint classes
INPUT	An individual of *Assembly* (Planetary_Gear_System_Assembly) contains the subassembly Planet_Gear_Holder, that in turn contains the Planetary_Gear_System_Assembly
OUTPUT	Planetary_Gear_System_Assembly belongs to both the classes *Assembly* and *NotAllowedAssembly*: an inconsistency is detected

7 Usage scenario of the product model in manufacturing

To demonstrate the application of the proposed model, let us consider a scenario of a distributed manufacturing facility where different manufacturing tasks like manufacturability evaluation, resource coordination, process planning, scheduling, fabrication, and logistics, have to be seamlessly integrated for product and process development. Here, the individual manufacturing tasks are modeled as functional software agents. To collaborate efficiently, these agents must be able to understand, communicate and negotiate for successful manufacturing tasks. This necessitates a need to formalize, encode and share manufacturing related knowledge. In other words, information represented must be semantically interoperable. Figure 10 presents an example scenario where collaboration takes place between a Design Mediator Agent (DMA), Manufacturing Evaluation Agent (MEA), Manufacturing Resource Agent (MRA) and possibly Other Manufacturing Agents (OMA). DMA is in charge of processing the job information. MEA is responsible for design evaluation and manufacturing best practices. MRA is responsible for resources (machines/ tools) allocation. For a detailed description of such an agent framework, refer to [21].

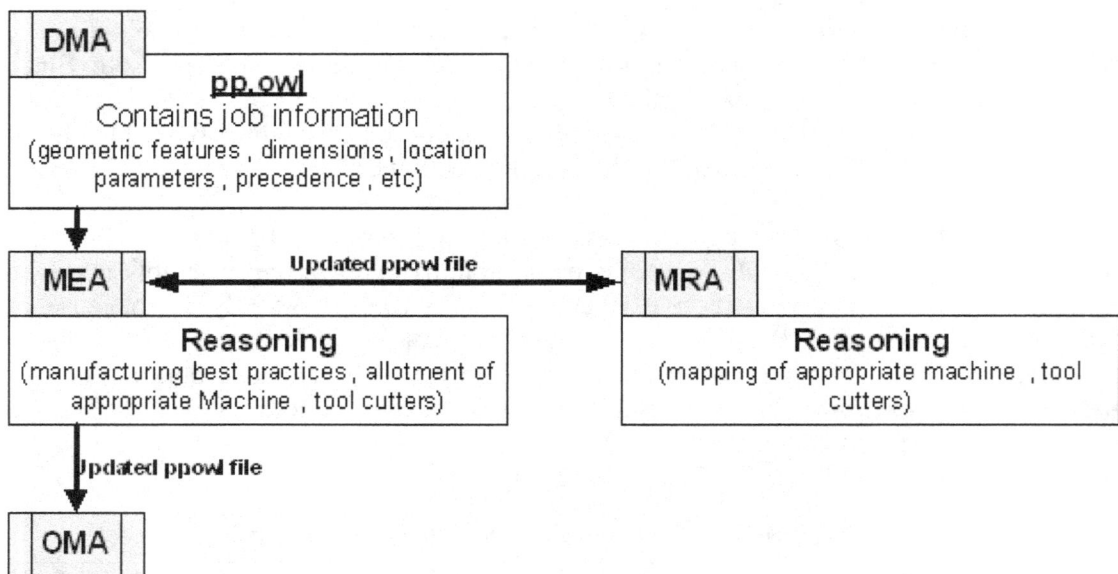

Figure 10: Manufacturing Information flow. Adopted/modified from [21]

Legend: DMA: Design Mediator Agent, MEA: Manufacturing Managing Agent, manufacturing Resource Agent, OMA: Other Manufacturing Agents

To process a manufacturing job we require information such as part features, naming conventions, location and dimensions, material specifications, associated manufacturing specific process information, including machine-tool information and associated tolerances. The proposed product ontology acts as a source for such product related data. This data, represented in an owl file, will be dynamically updated when exchanged between agents.

Figure 11: Prismatic part example

In this example scenario, consider the job to fabricate a simple prismatic part (pp), essentially a square boss with a hole in it as shown in the Figure 11. The instantiation of the product ontology specific to the prismatic part is represented in the pp.owl file. DMA processes the job request and sends the file (pp.owl) to the MEA for the purpose of manufacturability evaluation. An interpreter in MEA extracts the feature-related information from the pp.owl file and performs manufacturability evaluation. MEA supports inference-reasoning mechanisms that consider good manufacturing principles and machining processes for individual features [21]. Then MEA sends the results to the MRA for mapping of resources. MRA supports inference-reasoning mechanisms considering the availability of resources. Both the inference-reasoning mechanisms are based on description logic reasoning and rule-based reasoning.

Of the various feature level operations, we specifically consider the drilling operation to fabricate the hole. For implementation, note that in the pp.owl the hole in the prismatic part will be an individual of the class *Hole*, which is a subclass of the class *Feature* in the product ontology. For drilling, MEA considers the following manufacturing principle: for drilling a big hole with tight tolerances, a center drill is first used followed by a pilot drill. For a manufacturing drilling operation a center drill usually provides a starting hole for a larger-sized drill bit, like a pilot drill. To implement this principle, MEA must create a connection between the hole in the pp.owl and the sequence of drilling operations. Specifically, the DL expressivity will be useful to create the sequence of drilling operations while the rule-based reasoning mechanism will be useful to connect the operation sequence to the hole.

For creating the sequence of drilling operations, we use the patterns for sequences suggested in [22]. The detailed implementation in OWL-DL is represented in Figure 12.

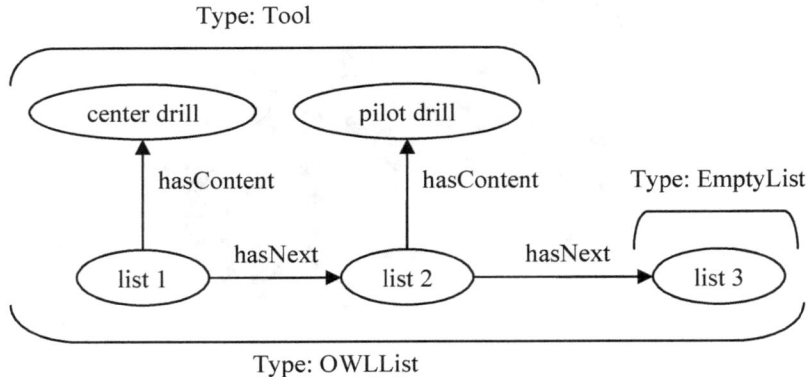

Figure 12: Sequence of drilling operations in OWL-DL

To implement the patterns we use specific DL expressivity (see Table 2) like:
- value restriction: OWLList can be connected through the property hasNext only to OWLList
- intersection of concepts: EmptyList is the intersection between OWLList and the concept of the individuals without the properties hasNext and hasContent.
- functional property: hasNext, hasContent
- cardinality restrictions: applied to EmptyList

Accordingly MEA will create (in the pp.owl) list1 and list2 as individuals of the class OWLList, list3 as individual of EmptyList, and pilot drill and center drill as individuals of the class Tool.

Now that the sequence is created, the connection to the hole is made through the following SWRL rule:
if (Hole hasdiameter Size bigger than x) and (Hole hasTolerance CilindricityTolerance less than z) → Hole isManufacturedThrough List1
hasDiameter and isManufacturedThrough are the properties of the Hole used to identify the diameter and the manufacturing operation respectively. After MEA executes the rule, it infers the corresponding operation sequence for the hole in the prismatic part.

Now, the updated pp.owl becomes available to the MRA for manufacturing resource allocation. The MRA will check if either the center drill or the pilot drill is not available. If one of those tools is currently used by a machine, the MRA creates an inconsistency in the pp.owl and sends the updated pp.owl to MEA for alternate solutions.

The following SWRL rule is executed in MRA:
if (center drill isUsedBy some Machine) → Hole isManufacturedTrough list3
The class Feature has cardinality 1 on the property isManufacturedThrough. Now, when the hole in pp.owl is connected to both list1 and list3, an inconsistency is generated.

When the MRA and MEA find a successful solution the results are updated in the pp.owl which becomes available to the OMA for further processing.

In this example scenario the content of the file pp.owl plays a central role. In the pp.owl file, description logic is used to represent the semantics of the product information, e.g., a value restriction is used to represent the connection between the hole feature in the prismatic part and the sequence of drilling operations. Description logic mechanisms and domain specific rules allow for reasoning on the product information. While using description logic mechanisms in this example, the consistency checking reasoning mechanism is specifically applied to investigate the availability of the resources to manufacture the hole. While using domain-specific rules, a SWRL rule is specifically applied to relate a manufacturing principle to that hole. This example suggests a future research direction on the role of ontological product models for semantic interoperability and reasoning in manufacturing-related processes.

8 Conclusions

To ensure interoperability of different systems and applications sharing product information across its different stages of lifecycle, a multiple view of the product model is required. Developing such a model is influenced by three factors: logical formalisms, computer interpretable languages and product information.

Usually logicians focus on logical theorem proving, computer scientists focus on theory of languages while product engineers focus on product information representation. The aim of this paper is to understand the issue that exists between these three research areas.

As a first step, we chose description logic as the logical formalism to represent product knowledge in a structured fashion. DL provides mechanisms for both explicit knowledge specification and implicit knowledge inference. These mechanisms, in the case of DL, are decidable, so that a model user can safely check and query the model. We choose OWL-DL language for implementing DL constructs in the product modeling domain.

As a second step, we chose domain-specific rules to increase the expressivity in the product model. Description logic and domain-specific rules are combined together to understand the level of expressivity required in product modeling. For the domain-specific rules, we chose SWRL, because it is compatible with the OWL editor Protégé. We then use OWL and SWRL to build product ontologies: Ontological Core Product Model for representing a generic product and Ontological Open Assembly Model to extend the previous one for representing mechanical assemblies. We use these ontologies to show how the expressivity of DL and domain-specific rules are used within the product modeling context. We finally use the planetary gear system example to instantiate the ontology. In this use case we show how the level of expressivity in the model allows both for explicit knowledge specification and implicit knowledge inference.

This paper outlined how to evaluate DL to capture both the information content and the abstraction principles with an aim of developing a consistent formal model for product assemblies. We believe this study can be extended to understand how to choose appropriate logical frameworks (OWL DL to OWL Full) for developing product ontology using OWL. Moreover, since OWL, the reasoners and the available tools are evolving we are constantly evaluating our approach.

9 Disclaimer

No approval or endorsement of any commercial product by NIST is intended or implied. Certain commercial equipment, instruments or materials are identified in this report to facilitate better understanding. Such identification does not imply recommendations or endorsement by NIST nor does it imply the materials or equipment identified are necessarily the best available for the purpose.

10 References

References

1. Taivalsaari, A., "On the notion of inheritance," *ACM Computing Surveys*, Vol. 28, No. 3, 1996, pp. 438-479.

2. Sudarsan, R., Baysal, M. M., Roy, U., Foufou, S., Bock, C., Fenves, S. J., Subrahmanian, E., Lyons K.W, and Sriram, R. D., "Information models for product representation: core and assembly models," *International Journal of Product Development*, Vol. 2, No. 3, 2005, pp. 207-235.

3. Artale, A., Franconi, E., Guarino, N., and Pazzi, L., "Part-whole relations in object-centered systems: an overview," *Data & Knowledge Engineering*, Vol. 20, No. 3, 1996, pp. 347-383.

4. Sudarsan, R., Subrahmanian, E., Bouras, A., Fenves, S., Foufou, S., and Sriram, R. D., "Information sharing and exchange in the context of product lifecycle management: Role of standards," *Computer-Aided Design*, Vol. to appear, 2008.

5. Ma, L., Mei, J., Pan, Y., Kulkarni, K., Fokoue, A., and Ranganathan, A.. Semantic Web Technologies and Data Management. http://www.w3.org/2007/03/RdfRDB/papers/ma.pdf . 2008.

6. Dolby, J., Fokoue, A., Kalyanpur, A., Kershenbaum, A., Schonberg, E., Srinivas, K., and Ma, L.. Scalable Semantic Retrieval Through Summarization and Refinement. http://domino.research.ibm.com/comm/research_projects.nsf/pages/iaa.index.html/$FILE/techReport2007.pdf . 2008.

7. Baader, F., Calavanese, D., McGuinnes, D., Nardi, D., and Patel-Schneider, *The description logic handbook*, Cambridge University Press 2003.

8. OMG. UML 2.0 Superstructure Specification. http://www.omg.org/cgi-bin/doc?ptc/03-08-02 . 2003.

9. Chen, P. P., "The Entity-Relationship Model: Toward a Unified View of Data," *ACM Transactions on Database Systems*, Vol. 1, No. 1, 1976, pp. 9-36.

10. Schenck, D., and Wilson, P. R., *Information modeling: the EXPRESS way*, Oxford University Press, New York, 1994.

11. Web Ontology Language (OWL). http://www.w3.org/2004/OWL/ . 2005.

12. Berardi, D., Cal, A., Calvanese, D., and De Giacomo, G.. Reasoning on UML Class Diagrams. http://www.dis.uniroma1.it/~degiacom/didattica/esslli03/ . 2003. Dipartimento di Informatica e Sistemistica, Università di Roma "La Sapienza".

13. Meta Object Facility (MOF) Specification. http://www.omg.org/docs/formal/02-04-03.pdf . 2002.

14. Protégé. http://protege.stanford.edu/ . 2008.

15. RacerPro. http://www.racer-systems.com/index.phtml . 2008.

16. SWRL. http://www.w3.org/Submission/SWRL/ . 2004.

17. Fiorentini, X., Gambino, I., Liang, V., Foufou, S., Rachuri, S., Bock, C., and Mahesh, M., "Towards an ontology for open assembly model," International Conference on Product Lifecycle Management 2007,2007, pp. 445-456.

18. W3C. Representing Specified Values in OWL: "value partitions" and "value sets". http://www.w3.org/TR/swbp-specified-values/ . 2005.

19. Liebig, T., "Reasoning with OWL: System Support and Insights," Computer Science Faculty, Ulm University, Technical report 2006-04, Sept. 2006.

20. Fenves, S., Foufou, S., Bock, C., Bouillon, N., and Sriram, R. D., "CPM2: A Revised Core Product Model for Representing Design Information ," National Institute of Standards and Technology, NISTIR 7185, Gaithersburg, MD 20899, USA, 2004.

21. Mahesh, M., Ong, S. K., Nee, A. Y. C., Fuh, J. Y. H., and Zhang, Y. F., "Towards A Generic Distributed and Collaborative Digital Manufacturing," *Robotics and Computer Integrated Manufacturing*, Vol. 23, No. 3, 2007, pp. 267-275.

22. Drummond, N., Rector, A., Stevens, R., Moulton, G., Horridge, M., Wang, H., and Seidenberg, J.. Putting OWL in order. http://ftp.informatik.rwth-aachen.de/Publications/CEUR-WS/Vol-216/ 216. 2006. OWLED '06 OWL: Experiences and Directions 2006.

www.ingramcontent.com/pod-product-compliance
Lightning Source LLC
Chambersburg PA
CBHW081811170526
45167CB00008B/3401